FOREST ANIMALS

Kizzi Roberts
M.S. Animal Science

Ways to use this book

Younger Readers

Read the book together and let your reader hear the clues and see the close-up picture as they guess the animal.

Older Readers

Read each clue aloud without showing the close-up picture. Let your reader try to guess the animal as you read the clues. Show the close-up picture if they need an extra clue.

Make It a Game!

Read the clues aloud and try to guess the animal in as few clues as possible. Come up with your own clues for each animal after reading the extra facts.

LEARNINGSPARKEDUCATIONALPUBLISHING

I eat **MEAT** and **PLANTS**.

I can climb **TREES**.

My fur can be **BLACK** or **BROWN**.

I **HIBERNATE** in winter.

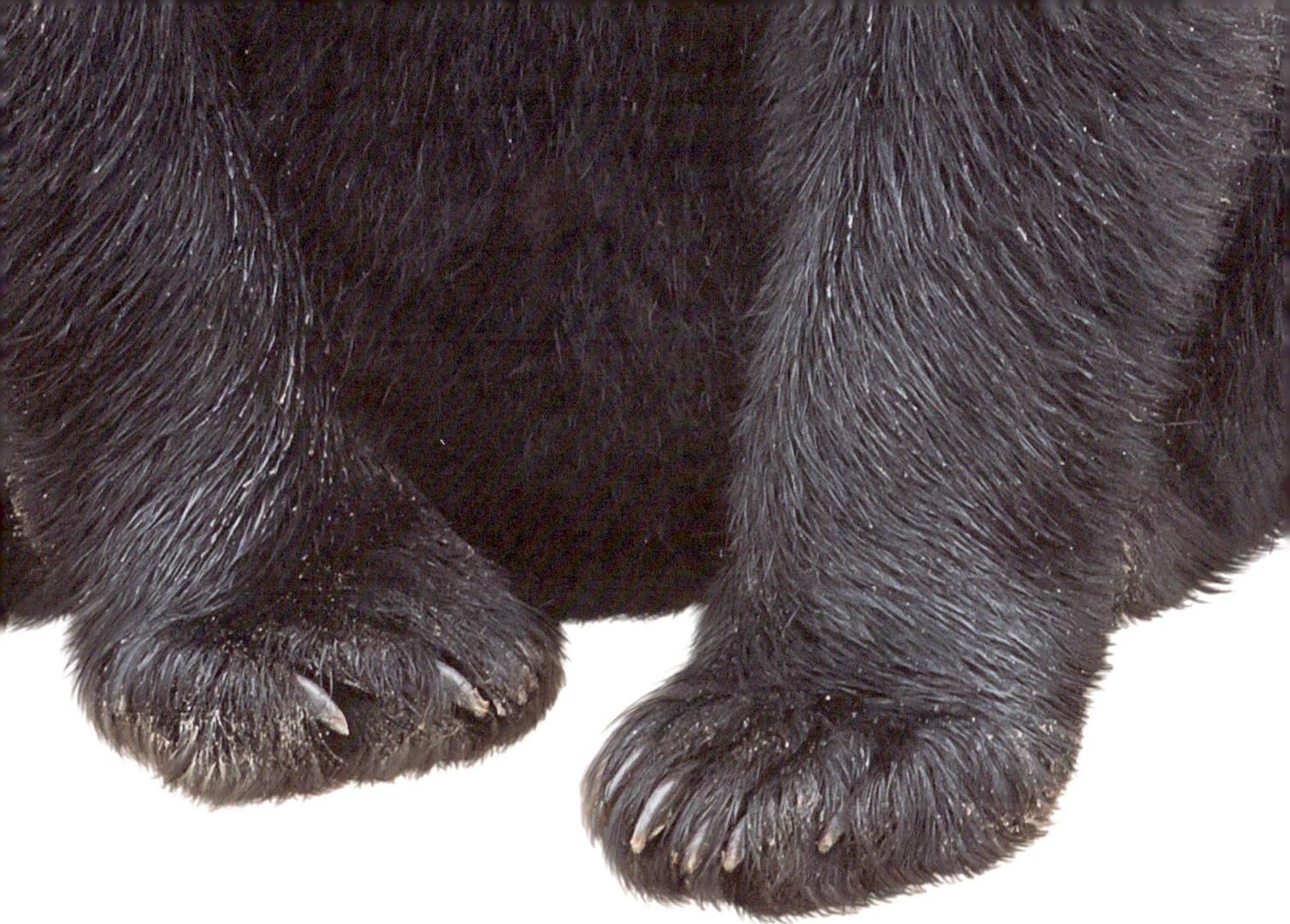

WHAT AM I?

I am a
BLACK BEAR.

BLACK BEAR FACTS

Black bears are omnivores. This means they eat both plants and animals. A black bear eats roots, berries, and insects. It also eats deer, elk, and even moose. Bears like to eat many different things. They also like human food. It's important for people to keep food secure when traveling or living near black bears.

Black bears are known for having black fur. However, some types of black bears have brown, rust colored, or even white fur. A black bear's thick fur lets it hibernate in the winter. During hibernation, a bear sleeps and does not need to eat. Female bears wake up to give birth during hibernation.

Black bears have short claws. These short claws help them climb trees. Black bears cannot retract their claws.

I am **SMALL**.

I eat both **PLANTS** and **ANIMALS**.

I burrow **UNDERGROUND**.

I can store food in my **CHEEKS**.

WHAT AM I?

I am a **CHIPMUNK.**

CHIPMUNK FACTS

• • • • • • • • • • • • • • • • • •

Chipmunks are omnivores. They eat plants, mushrooms, and seeds. They also eat bird eggs, small birds, worms, and insects. Chipmunks stuff food in their cheeks and carry it back to their burrow to store. The skin stretches up to three times the size of the chipmunk's head.

Chipmunks are good at climbing trees, but they live in burrows underground. A chipmunk's burrow has rooms for storing food and sleeping. A chipmunk can dig a burrow up to three feet deep and thirty feet long. While chipmunks don't hibernate, they do spend a lot of time sleeping in their burrows during winter.

Chipmunks are small. They can grow up to twelve inches long. Despite their small size, they are known for making loud chirping sounds. The chipmunk was named for the "chip, chip" sound it makes.

I am known to be **CUNNING** and **SMART**.

I have excellent **HEARING**.

I like to eat **RODENTS** and **RABBITS**.

I make a **DEN** for my young.

WHAT AM I?

I am a **RED FOX**.

RED FOX FACTS

• • • • • • • • • • • • • • • • •

The red fox is known for its red fur and white-tipped tail. Sometimes a red fox can have large patches of gray fur and might be confused with a gray fox. However, the gray fox has a black-tipped tail. The red fox has four black paws and black-tipped ears. The red fox has large ears and excellent hearing. It can even hear rodents underground!

The red fox likes to eat rabbits and other rodents. If food is scarce, it will eat a variety of other things. It is good at finding food. Its ability to find food gives it a reputation for being smart and cunning.

The red fox is also smart when building dens. The female builds multiple dens to help protect against predators. If something happens to one den, she can move her babies into a new den. A den can be up to 75 feet long with rooms for sleeping and storing food. The male and female fox raise their babies together.

I am mostly active at **NIGHT**.

I like to dig to find **FOOD**.

I use a **DEN** all year long.

I have a **STINKY** spray to defend myself.

WHAT
AM I?

I am a **SKUNK.**

SKUNK FACTS

• •

The skunk is well known for being black and white with a stripe down its back. Some skunks have a single stripe, others have two stripes. Some types of skunks have white spots or broken stripes. While most skunks are black and white, some species may be gray or brown.

The skunk is also known for the stinky spray it makes. When scared or attacked, a skunk sprays the oily, smelly substance from glands under its tail. The substance resists water, making the smell difficult to remove. A skunk has a mild temper and usually only sprays after hissing and stomping its feet.

Skunks have strong front feet. Their feet are good for digging dens and digging for food. Skunks use dens year round. A skunk sleeps in its den during the day and is active at night. A skunk will stay in its den longer during the winter, but it does not hibernate.

I use **SOUND** and **SMELL** to find food.

I eat mostly **INSECTS**.

I find most of my food in dead or dying **TREES**.

I use my beak for **DRILLING** and **DRUMMING**.

WHAT AM I?

I am a **WOODPECKER**.

WOODPECKER FACTS

• • • • • • • • • • • • • • • • •

Woodpeckers are often heard before they are seen. A woodpecker uses its strong beak to drill into trees in search of food. The sound of a woodpecker drilling is easily identified. Woodpeckers also use their beaks to make holes for nests in trees. They produce a drumming sound which attracts a mate.

Woodpeckers listen for insects inside trees before they start drilling. Some insects produce a scent that woodpeckers can smell. Woodpeckers mostly find food inside dead trees. After a woodpecker drills a hole, it uses its long tongue to reach into the crevice and find food. A woodpecker's tongue can measure as long as a third of its body length.

Woodpeckers live almost everywhere. They can be small or big depending on the species. Small woodpeckers make small holes in trees. Large woodpeckers make bigger holes in trees.

I am usually **NERVOUS** and **SHY**.

I eat **PLANTS**.

I bound away quickly
to escape **PREDATORS**.

I wave my tail like a
FLAG when **STARTLED**.

WHAT AM I?

I am a
WHITE-TAILED DEER.

WHITE-TAILED DEER FACTS

White-tailed deer have reddish-brown coats in summer. In winter, the coat fades to more of a gray color. Only male white-tailed deer have antlers. Antlers fall off every year in the winter. New antlers grow every spring. White-tailed deer get their name because the underside of their tail is white. When they are startled, they wave their tail like a flag as they bound away.

White-tailed deer are nervous and shy. They are easily startled and bound away quickly when frightened. They can reach speeds up to 30 miles per hour. White-tailed deer are also strong swimmers. They may swim across a river to get away from predators.

White-tailed deer eat only plants. They usually eat before dawn and at dusk. During the middle of the day they are not very active.

I eat **MEAT**.

I enjoy **SWIMMING** and **CLIMBING** trees.

I can **LEAP** up to ten feet in the **AIR**.

I hunt at **NIGHT**.

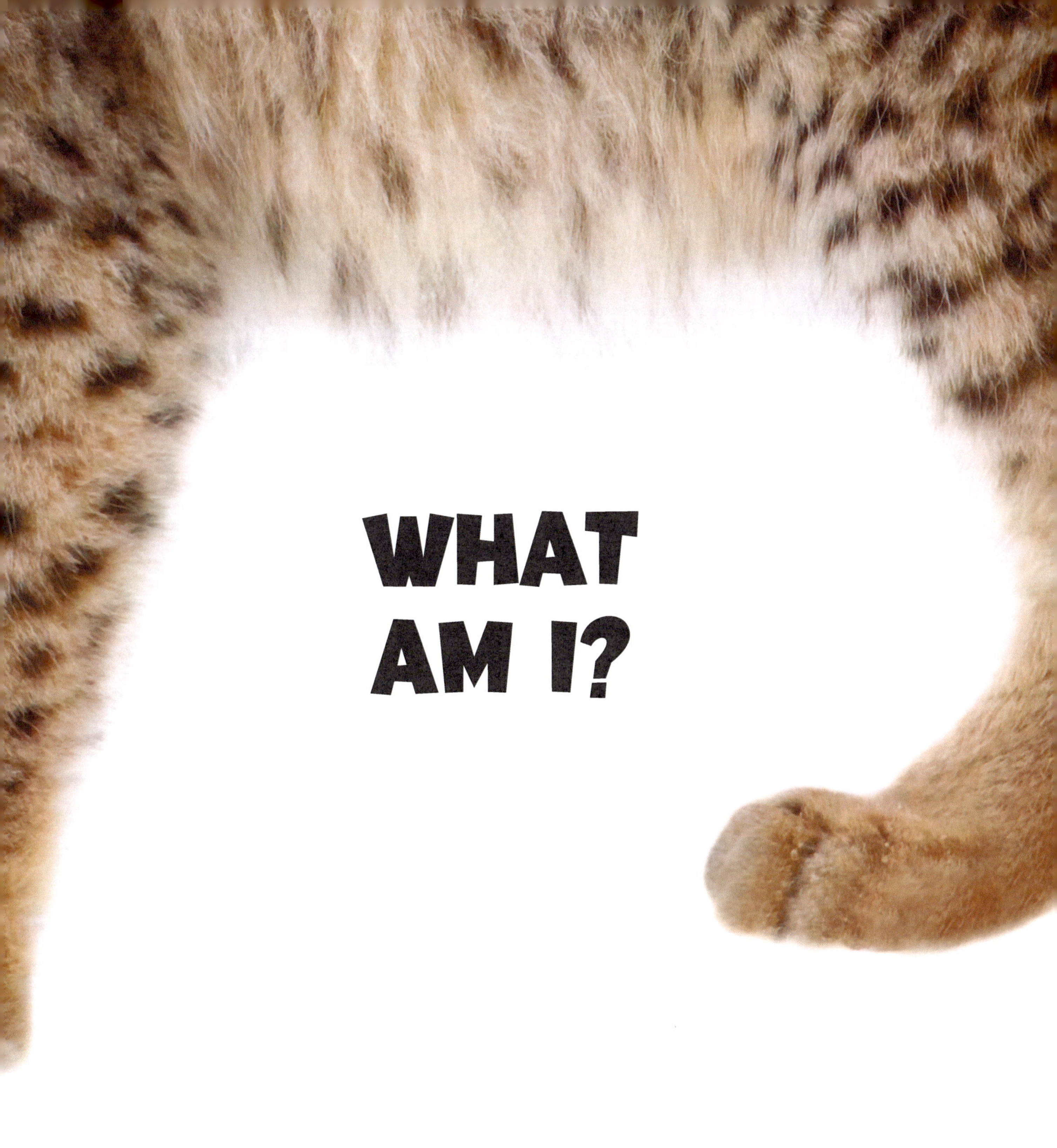
WHAT AM I?

I am a **BOBCAT.**

BOBCAT FACTS

Bobcats have short tufts of fur on top of their ears. A bobcat also has a short tail. The bobcat is sometimes confused with other lynx species. The bobcat is smaller than other kinds of lynx and lives in warmer areas.

Bobcats are carnivores. This means they only eat meat. Bobcats mostly eat rabbits, but they eat other small mammals if rabbits are scarce. In winter, a bobcat might go after larger prey, like deer. Bobcats can also jump up to 10 feet in the air. Sometimes a bobcat jumps and catches a low-flying bird. Bobcats only hunt at night.

Bobcats are excellent climbers. They climb trees and rocky terrain. Bobcats are also good at swimming.

Clues by YOU!

It's your turn to create clues. Look at the pictures and read the facts for the following forest animals. Can you come up with four clues for each animal based on their appearance, behavior, or some other fact?

Tell your clues to a friend or family member and see if they can guess the animal without looking at the picture.

I am a WILD TURKEY.

The wild turkey is one of the most recognizable birds in North America. Males grow up to 25 pounds and are known for their long beards. The beard grows from its chest, and sometimes female turkeys grow beards too. Male and female turkeys have different shaped poop. Males have J-shaped, while females have spiral shaped poop.

The wild turkey runs up to 18 miles per hour. It flies up to 60 miles per hour. Turkeys see three times better than perfect human vision.

I am a SQUIRREL.

Squirrels are known for their amazing tree-climbing skills and ability to store nuts. A squirrel buries up to 10,000 nuts in one season. Squirrels find buried nuts using their sense of smell. Some species of squirrels can smell buried food under one foot of snow. Different species of squirrels have brown, black, red, or gray fur. All squirrels are rodents.

I am an OPOSSUM.

The opossum is the only marsupial in North America. Marsupials have pouches (like kangaroos) for their babies. The opossum has a prehensile tail. This means it uses its tail to grab things like branches.

The opossum is active at night and eats bugs and insects. It also eats mice and rats. The opossum is known for "playing possum", or playing dead, when threatened.

Text copyright © 2024 Kizzi Roberts

Photographs © peterwey/depositphotos.com; nicolaselowe/depositphotos.com; Rajen1980/depositphotos.com; vaclavmatous/depositphotos.com; twildlife/depositphotos.com; JimCumming/depositphotos.com; deepspacedave/depositphotos.com; Farinosa/depositphotos.com; kwasny222/depositphotos.com; pictureguy/depositphotos.com; NynkevanHolten/depositphotos.com; OndrejProsicky/depositphotos.com; dimjul/depositphotos.com; igordabari/depositphotos.com; jill@ghostbear.org/depositphotos.com; svetas/depositphotos.com

Published in March 2024 by Learning Spark Educational Publishing in Rogersville, Missouri. Learning Spark Educational Publishing is an imprint of Elemental Ink Publishing LLC.

Library of Congress Control Number: 2024905927

Hardcover: 979-8-88884-030-6; Paperback: 979-8-88884-029-0
Edited by Carrie Rodell. Book design and layout by Kizzi Roberts.

www.LearningSpark.com

www.ingramcontent.com/pod-product-compliance
Lightning Source LLC
Chambersburg PA
CBRC090805300326
41914CB00069B/1640